SANTA CRUZ
DEL VALLE DE LOS CAÍDOS

Editorial Everest, S. A., wish to thank all members of the National Heritage who have so kindly collaborated in the production of this book.

Photography: Francisco Díez
Oronoz

Layout: María Casas

Ctra. León-La Coruña, km 5 - LEÓN
ISBN: 84-241-4967-X
Legal deposit: LE. 398-1990
Printed in Spain

EDITORIAL EVERGRÁFICAS, S. A.
Ctra. León-La Coruña, km 5
LEÓN (Spain)

SANTA CRUZ
DEL VALLE DE LOS CAÍDOS

JOSÉ MANUEL TORNERO

EDITORIAL EVEREST, S. A.

MADRID • LEON • BARCELONA • SEVILLA • GRANADA • VALENCIA
ZARAGOZA • LAS PALMAS DE GRAN CANARIA • LA CORUÑA
PALMA DE MALLORCA • ALICANTE – MEXICO • BUENOS AIRES

LOCATION AND HISTORY OF THE CONSTRUCTION OF THE HOLY CROSS

The Holy Cross in Los Caídos Valley (The Valley of the Fallen Heroes) is a magnificent feat of engineering and architecture which, combined in unique harmony with nature, houses a sanctuary in the heart of the rock itself.

Situated within the municipal district of El Escorial, close to the capital of Spain, the area comprises 1,365 hectares and is bounded by the town of Guadarrama, a small river called the Gautel, La Solana country estate and the Jurisdicción Highlands.

It is at a height of between 985 and the 1,758 metres of Mount Abantos, 57 kilometres from Madrid and seven from El Escorial, within the Guadarrama mountain range, in the place called Cuelgamuros.

Between the Leones mountain pass and Mount Abantos the valley contains scenery of superb beauty and great personality, the main feature being stone, which alternates with groups of trees such as pines, oaks and poplars, and clumps of plants such as thyme, rock roses and daphnes. At the far end of the gorge, in isolated majesty, rises the stony bulk of Risco de la Nava rock.

In the middle of this unique rock form stands, majestically, the monument; it does not break with the surroundings, it offers no violent contrast, but like a whimsical product in rock made by the elements it is a Cyclops-like mass evoking hidden times.

The place was chosen by General Francisco Franco, the previous head of State; his also was the idea and concept of the monument which would be both a great church housed within the rock and a monumental cross crowning it, as well as that of the Benedictine monastery which would take care of the monument.

The express desire of the founder was that of constructing the final dwelling place for soldiers from both sides who had fallen during the civil war of 1936-1939. There are chapels with the remains of 40 000 victims, most of whom belonged to General Franco's army. Both Franco himself and José Antonio Primo de Rivera lie to rest beside the high altar of the basilica.

The construction of the monument was

decreed by Act of Parliament on 1st April 1940. It was to be directed by Pedro Muguruza Otaño, the Director General of Architecture, with the assistance of Oyarzábal and Muñoz Salvador.

The work was quickly begun, and the necessary geological and climatic analyses were carried out. The work on the present-day Hospice and Centre of Studies was begun, and the mountain began to be perforated, but Pedro Muguruza fell ill, and in 1950 the direction of the undertaking had to be handed over to Diego Méndez González. The work continued, but decisions about two fundamental parts of the building — the crypt and the cross — were still to be taken.

The main problem was the construction of the basilica, which implied the perforation of Risco de la Nava rock so that the building could be housed in it. Also, the express magnificence of the project called for a greater perforation of the rock than that which had been carried out by that time, as it was merely a tunnel measuring 11 metres by 11 metres. These dimensions have been maintained in the area leading to the present-day main nave. The nave itself now reaches 22 metres. A further problem was the enormous, heavy monument-

1. *Los Juanelos on the way to the Holy Cross.*

al cross which was to be placed on the hollowed-out section.

This accumulation of circumstances made the process of construction arduous and complex, as no decision could be taken without its having been profoundly considered beforehand. In 1950, the work on the present-day residence was finished; in 1952 plans were made for the forecourt and the expansion of the crypt was begun; work on the crypt continued through 1953 and 1954; in 1954 plans were made for the completion of the cross. In 1955 most of the work was concentrated on the vault in the crypt and in 1956 the choir was completed, amongst other things. Finally, in 1957, plans were made for the monastery, the novitiate and other remaining parts. The building work of the monument was finalized in 1958.

2. *Entrance to the valley.*

THE APPROACH AND THE FORECOURT

Once in the valley, a wide road leads to the monument itself. Watching over the approach, like guardians unchanged throughout time, are the famous *Juanelos,* on either side of the road. They are four monolithic stone posts, carved in the 16th century to be used by the famous Juanelo Turiano; they come from the fine quarries in Nambroca and Fonseca.

Finally, a flight of steps leads up from the road to the wide forecourt. These steps are made up of two sections of ten, the number symbolising the Ten Commandments given to Moses and which continually appears in different parts of the monument.

The forecourt covers an area of 30,600 square metres, and is made up of three quite different parts: a central part, limited by a distinct handrail, and two side parts,

3. *Detail from the entrance to the crypt.*

4. *The large forecourt, made up of three parts. Aerial view.* ▼

5. *The Holy Cross: view from the Gardens.* ▶

which are lower than the central one and communicate with it by flights of steps.

A further flight of steps leads to the entrance to the crypt, which is flanked by two magnificent rows of very sober arches. This façade acts as a great curtain hanging before the lower strip of rock formation. At the same time its very solidness, broken only by the round archways, affords it a certain harmony with the stone mass is hides.

The monument as a whole, with its three basic elements — this lower façade, the rock and the cross — takes on an almost pyramidal shape which is then more clearly repeated in one of its components: the cross. This wise choice of shape is to be praised, as the pyramid has a form which originates ascensional dynamics of greatest effect; it is not only suitable for this work but also exemplifies the religious and spiritual character of the monument. The real believer grows in his Faith and advances on his road towards the Truth in the same way as there is here an ascent from the depths of the earth towards the immensity of heaven.

THE CROSS

If the soul of the monument is inside the basilica, on the outside the cross is its crucial element. Placed on the rock itself, it seems to be a prolongation of it, as if it were a game being played by nature, as if the elements had capriciously fashioned it. Its profile against the sky affords it a solemnity which overawes any visitor who contemplates it. Its size strikes you, envelops you; once in the valley the sight of it in the distance progressively penetrates the mind until at the foot of it its attraction cannot be escaped.

Made of concrete and granite, the cross comprises three basic elements. Rising up out of the rock is the lower part — a strong base with statues of the Evangelists on the four corners. The second smaller section rests on the first. Here are placed figures of the four Cardinal Virtues; here begins the vertical section of the cross. The final section is the cross itself.

The dimensions are gigantic —the lower section measures 25 metres, the upper section 42 metres and the cross a total of 150 metres from the base and 300 from the forecourt. The cross itself is formed by the interpenetration of two rectangular prisms which make up the transversal section of a Greek cross. Finally, a slightly raised ogee completes the whole.

The construction was carried out from the inside; in order to facilitate the work of transporting the materials a hole was tunnelled through from the back of Risco de la Nava rock as far as the vertical at the base of the rock.

As the interior is hollow a lift has been installed; the arms of the cross, which are 46 metres in length, contain very wide passageways.

Access to the base is by a funicular railway which has been cut into the rock in a south-east to north-west direction, so that it cannot be seen from the forecourt, thus maintaining the aesthetic value of the monument and the scenery around it.

At first it had been decided to decorate the monument with statues of the twelve Apostles, but this idea finally resulted in those of the four Evangelists and the four Cardinal Virtues. These figures have been conceived as a transition from the ridge of rock at the foot of the cross itself, so that the monument could maintain the fluid, uninterrupted pyramidal line mentioned above.

The sculptural aspect of the monument produced as considerable difficulties as the architectural aspect, due to the great size of the figures and the problems in applying the blocks of stone forming the groups. The conditions made it necessary to discard any idea of using normal criteria, using instead gigantic proportions in harmony with the form of the valley and the colossal rocks around the cross. In this way no violent contrast between the rock and the shaft of the cross can be appreciated, as they appear to be gently united by these huge statues.

The statue of St. John seems to be bent forward over its eagle, as if captured at one precise moment, like in a photograph. St. Luke is astride a bull's head, carrying the Gospel in his hand; St. Mark is very twisted, baroque in style, with his symbolic

6. *The Holy Cross.*▶

7. *Detail of the funicular railway.*

8. *The base of the cross: statues of the Evangelists and the Virtues.* ▶

9. *The funicular railway, the line from the forecourt. General view.*

10. *St. Mark.*

11. *St. John.*

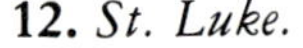

12. *St. Luke.*

13. *St. Matthew.*

14 and 15. *Statues on the cross: details of the Virtues.*

animal, the lion; St. Matthew appears very calm, bent over, reading.

Each evangelist is 18 metres high; the total weight of this group of statues is 20,000 tons, which added to the 181,740 the cross weighs makes a total of 201,740 tons.

The statues, by Juan de Ávalos in black stone from Calatorao, clearly denote the hard, triumphal style of this sculptor; they are monumental and overawe the spectator, who with his own modest size feels surpassed by such masses of stone.

While the figures of the evangelists, at the heavy base, are completely outsize compared with their setting, so that they overcome it, the statues of the four Cardinal Virtues hardly meet theirs, which are of lesser dimensions. In this way the figures aid in the formation of the pyramidal tendency of the cross, which has already been mentioned under the *architectural* aspects, emphasizing the volume of the lower part and slightly restricting itself in the upper part. In the same way the virtues are much more elongated then the evangelists; they have proportions which are more rounded but less stylised.

THE DOORWAY

Almost excessively simple, the doorway is composed of a double round arch within a completely plain framework. On either side is a row of round archways completing the whole; the curved form surrounds the visitor, inviting him to enter within the confines of the mountain and discover the treasures it contains.

Decoration is almost non-existent, being the embossed work on the inside arch of the doorway, the Pietà crowning the work and the doors giving access to the interior.

These doors, made of bronze, are by the sculptor Fernando Cruz Solís; the decoration on them is made up of consoles showing the fifteen mysteries of the rosary and a group of apostles.

The round arch shows the first, second and fourth of the glorious mysteries: the Resurrection, the Ascension of Christ and the Assumption of the Virgin; the jambs are decorated in four parts, the first three made up of four sections with the rest of the mysteries and the fourth with the group of apostles. The mysteries of the rosary appear in the following order: the Annunciation, the Descent of the Holy Spirit, the Coronation of the Virgin, the Prayer in the Garden, the Visitation, the Birth of Christ, the Flagelation, the Crown of Thorns, the Circumcision, Christ the Child in the Temple, the Road to Calvary, and the Crucifixion and Death of Christ.

▲
▲ **16.** *The doorway, general view.*

▲ **17.** *The doorway, detail of one of the side arcades.*

19. *The entrance doors, general view.*

► **18.** *The entrance doors. Detail of the tympanum: the Resurrection, the Ascension. From the two first sections: the Annunciation, the Descent of the Holy Spirit, the Coronation of the Virgin and the Prayer in the Garden.*

▲

▲ **20.** *The entrance doors: detail of the third and fifth sections — the Visitation, the Birth of Christ, the Circumcision, and Christ the Child in the Temple.*

▲ **21.** *The entrance doors: detail of the fourth and sixth sections — the Flagellation, the Crown of Thorns, the Road to Calvary and the Crucifixion and Death.*

22 and 23. *The entrance doors: the seventh section — detail of the Apostles.*

24 and 25. *The entrance doors: the eighth section — detail of the Apostles.*

26. *The Pietà, on the cornice of the main doorway.*

This relief work, which shows very good use of light and dark, occupies almost the whole surface of the sections containing it, and in only a few of them can the background be seen — in some cases architectural and in others scenic. Of interest is the form of the bodies which is very varied, with dynamic, vigorous bodies next to other more deliberate and languid ones, as well as the excellent study of the drapery, in which a clear geometrical tendency in the hanging can be appreciated.

Above the doorway the cornice completes the whole, with the Pietà in which Juan de Ávalos again demonstrates his skill as a sculptor. Christ appears lying in the warm arms of His mother, whose touching expression shows her sorrow at the death of her son. Worthy of attention is the care given to the schulpting of the body of Christ, as well as the natural position of His body, with one arm falling back because of its own weight. Also of interest is Mary's soft complexion, the result of effective smoothing.

This group, in black stone from Calatorao, measures 12 metres long by 5 metres high, and is in stark contrast with the simplicity of the rest of the doorway, which acts as an impressive pedestal for the scene above it. The Pietà is in fact the centre of the façade as such, and it draws the attention of all those who are about to enter the crypt. The arcades on either side, mentioned above, merely serve to underline the Pietà.

THE CRYPT

In the crypt all the dificulties in construction which the project implied appeared; the different solutions which were found had to be very carefully studied, so that even nowadays the plan carried out to enable this magnificent work to be realised amazes many experts in the field. The major problem was that not only did vertical pressure have to be faced (which is normal in any construction), but also lateral pressure.

The total length is 262 metres; the maximum height, in the transept, in 41 metres. It is made up of a series of sections forming the porch, the hall, the middle section, the large nave and the transept. Although each section has its own characteristics and specific decoration, which increases progressively from the entrance porch to the transept, they are all determined by their function as part of a physical and spiritual whole.

27. *The transept of the crypt: one of the statues representing the Armed Forces.*

The porch is made up of four large pillars each supporting transverse arches and vaults with lunettes corresponding to the lateral arches. It is the simplest and most austere of the sections named, as it has no decorative elements apart from architectural ones.

The hall is more decorated, with large inclining pillars, transverse arches and a vault with very austere coffers, but it contains no outstanding sculptured or specific decorative elements of any kind.

A change of style can be appreciated in the middle section, with its growing vault. Two large recesses house statues of enormous archangels by Carlos Ferreira. Standing on guard, with their wings outspread, their arms resting on swords, they watch over the nearby entrance to the nave. Here the layout is far more schematic than in the previous work by Ávalos or in the bronze entrance doors. The details disappear, as in the case of the faces of these angels, which show very little expression. This middle section really acts as an area of transition, not only in the decorative aspect (it is far more adorned) but also in the architectural aspect. It acts as a hallway to the large nave, leading to and preparing for the main part of the work.

After a stairway, again with the symbolic

28 and 29. *The middle section of the crypt. The Archangels Gabriel and Michael.*

ten steps, comes the grille before the nave of the crypt. By José Espinos Alonso, it is made up of three parts which are divided into sections by four pillars, two at the side and two others in the centre which act as jambs for the central door. Saints, martyrs and the four Evangelists are represented on the pillars. These saints again appear on the cupola and are of great symbolic significance with regard to the idea of the monument.

St. John of the Cross, St. Maurice, St. Stephen, St. Barbara, St. Francis of Assisi and Joan of Arc are among the 40 different figures on the pillars; 20 of them look towards the hallway, the other 20 towards the high altar. They are placed in small round vaulted niches, carefully cut into the

30. *The crypt, with the grille before the nave. General view.*

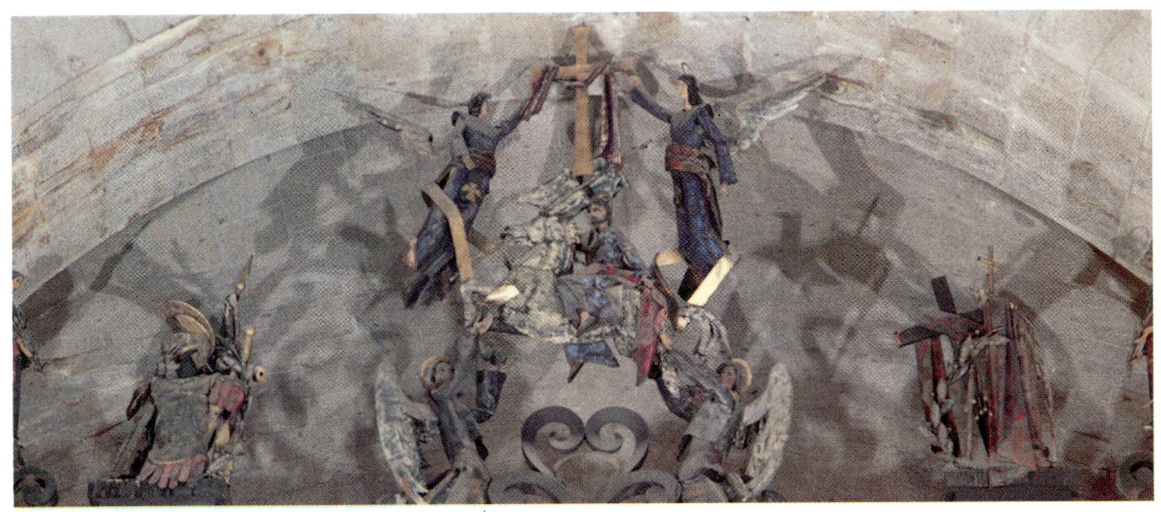

31. *The grille before the nave: detail of the coronation.*

32 and 33. *The crypt: details of the grille. St. Domingo de Guzmán and St. Ferdinand.*

pillars. Below each figure is a small name-plate in relief work with the name of the saint represented.

Finally, crowning the work, is a sculpted cresting, made up of angels and emblems of saints and heroes, with the figure of St. James in the centre, crowned with a cross and angels. The light paintwork on the figures and the numerous gilt details upon the grille offer a delicate note of colour in the austerity of the monument.

Beyond the grille is the section that comprises the large nave. It is in four sections, divided by a series of transverse arches forming coffers in the vault which show the rock itself. Six small recesses, or

34. *The crypt: detail of the large candle holders.*

chapels, are cut into the walls; each of them contains an alabaster relief with an altar dedicated to the Virgin Mary. Our Lady of the Immaculate Conception is followed by Our Lady of Carmen and Our Lady of Loretto, these being the patrons of the Army, the Navy and the Air Force. The first two are by Carlos Ferreira and the third by Ramón Mateu. Then comes Our Lady of Africa, followed by Our Lady of Mercy and Our Lady of the Pillar, by Ferreira, Lapayese and Mateu respectively.

These reliefs are half-way between high relief and statues, and although closer to the former they do exceed the usual volume.

This volume seems greater as the back of the recesses has not been worked, but has been left bare, so that the figures of the Virgins appear to stand out more. The soft, light modelling together with the material used — alabaster — give these figures a serene, peaceful air, which is extremely fitting in this nave.

As a complement the chapels contain altar fronts with pictures of the Virgin and triptyches painted on leather, like Mediaeval Spanish embossed leather, as well as statues of the Apostles by Lapayese and his son. Without breaking the dominant austerity of the nave, the golden reflections of the frontals and the warm colours of the triptyches offer a marked note of colour which attracts the attention of the visitor.

◀ **35.** *The crypt: the chapel of the Immaculate Conception. Detail of the alabaster relief over the entrance to the chapel.*

▲ **36.** *The chapel of the Immaculate Conception: detail of the relief decorating the altar front.*

▲ **37.** *The chapel of the Immaculate Conception: general view of the interior.*

▼ **38.** *The chapel of the Immaculate Conception: statue of James the Apostle.*

39. *The crypt: general view of the nave from the entrance to the far end.* ▶

▼ **40.** *The crypt, the chapel of Our Lady of Carmen: detail of the alabaster relief over the entrance to the chapel.*

▲ **41.** *The chapel of Our Lady of Carmen: statue of John the Apostle.*

◀ **42.** *The chapel of Our Lady of Carmen: general view of the interior.*

43. *The crypt: the chapel of Our Lady of Loretto: detail of the alabaster relief over the entrance to the chapel.*

44. *The chapel of Our Lady of Loretto: general view of the interior.*

▼ **45.** *The crypt: the chapel of Our Lady of Africa: detail of the alabaster relief over the entrance to the chapel.*

▲ **46.** *The chapel of Our Lady of Africa: detail of the triptych.*

◀ **47.** *The chapel of Our Lady of Africa: general view of the interior.*

▼ **48.** *The chapel of Our Lady of Africa: detail of the relief decorating the altar front.*

49. *The crypt: the chapel of Our Lady of Mercy: detail of the alabaster relief over the entrance.* ▼

50. *The chapel of Our Lady of Mercy: detail of the triptych.* ▼

51. *The chapel of Our Lady of Mercy: general view of the interior.* ▶

52. *The crypt: the chapel of Our Lady of the Pillar: detail of the alabaster relief over the entrance to the chapel.*

53. *The chapel of Our Lady of the Pillar: general view of the interior.*

54. *Detail of the tapestry of Hanging III.*

Hanging I: The Revelations made to St. John when on Patmos.
Hanging II: The Final Judgement.
Hanging III: The destruction of humanity by plagues and the adoration of the lamb.
Hanging IV: Enoch and Elijah.
Hanging V: The fight between the angels and the devils trying to attack the woman dressed as the sun.
Hanging VI: Triumph of the Gospel.
Hanging VII: The wedding of the Lamb.
Hanging VIII: The triumph of the Church over the devil.

THE APOCALYPSE TAPESTRIES

These chapels have distributed among them large murals made up of eight magnificent tapestries based on St. John's *Apocalypse.* Made by Guillermo Pannemaker from Brussels, whose mark appears on four of them, they are a collection of marvellous hangings made not only in wool and silk, but also in gold and silver.

Dating from before 1540 they were acquired by Felipe II and brought to Spain in 1553. Unfortunately, the name of the original designer is not known, but evidence suggests that some designs are by Durero and Juan de Brujas, and it is supposed that Bernardo Van Orley was probably the artist responsible for this original interpretation, in which the then contemporary Flemish style has been united with the new Italian tendencies.

Also of interest are the borders of the hangings, a true example of detail. There is a wide band in the lower part with texts alluding to the theme expressed; the rest of the tapestry is filled with highly exuberant floral designs within which several exotic birds can be seen.

The original tapestries now no longer hang in this basilica in Los Caídos Valley (the Valley of the Fallen Heroes). They have been replaced by magnificent copies, made with the same materials as the original ones and cleverly aged. It took more than ten years to make them.

These hangings recount the texts of St. John in chronological order, including the apotheosis of the Lord, who from His glory contemplates heaven.

55. *Tapestry I: The revelations made to St. John when on Patmos.*

IÆ. CAPIET. MERCEDEM. ECCLESIA. VICTRIX
ELICA. PERPETVO. REGNA. TENEBIT. OVANS.

56. *Tapestry II: The beginning of the Final Judgement.*

ANT. FORTES. PROPRIO. QVOS. SAGVINE.
ET. A. RECTO. TRAM TE. TVRBO. TRAHAT.

57. *Tapestry III: The destruction of humanity by plagues and the adoration of the Lamb.*

SANCTA.ERRORES.VARIOS.DOCTRINA.REPELLIT
ERVS.AMOR.FIDIS.MENTIBVS.INSERITVR

58. *Tapestry IV: The story of Enoch and Elijah.*

59. *Tapestry V: The fight between angels and devils trying to attack the woman dressed as the sun.*

60. *Tapestry VI: The triumph of the Gospel.*

61. *Tapestry VII: The wedding feast of the Lamb.*

62. *Tapestry VIII: The triumph of the Church over the devil.*

NS. MONSTRAT. QVI. QVONDAM. ECCLESIA. CREVIT
QVO. QVE. VIGENS. OLIM. FASCE. PREMENDA. MANET

63. *General view of the transept.*

THE TRANSEPT

Another flight of ten steps leads to the following section, that of the transept, containing statues representing the Armed Forces by Luis Antonio Sanguino and Antonio Martín. Prominent in these statues is the contrast between the coarse, wrinkled clothing and the softly polished arms and faces.

The dominant classicism of this section is broken only by the four main arches supporting the cupola; this contains raised wedge stones, affording a marked widening.

At the end, closing the perspective, is the choir. Semi-circular in shape and on three levels, with a total of 70 stalls, it is in very simple classic style. The choir is in wood, except for some details such as the

64. *The Archangel Gabriel.*

65. *The Archangel Michael.*

66. *Azrael.*

67. *St. Raphael.*

statues at the top, which are in gilded bronze. The backs of the stalls are in fine relief work, a result of the long Spanish tradition in the construction and decoration of choirs; this area is divided off from the rest of the basilica by a carefully-worked low grille.

Two chapels are situated in the side arms: the chapel of the Most Holy and the chapel of the Way of the Cross. The latter is the more impressive. It contains, below a large cross, a beautifully-made Calvary by Lapayese. Christ recumbent, in fine alabaster, contrasts greatly with the surrounding figures of St. John and the Virgin, in ostentatious colours.

In the centre of the transept, precisely below the monumental cross, is the High Altar, in finely polished granite. The front of the altar is decorated with a bas-relief in gold plate showing the Burial of Christ. Designed by Diego Méndez it was made by Espinós. The back of the altar shows a bas-relief in gold plate of The Last Supper.

But the finest piece is the Christ crucified above the altar made by Beovide and painted by Ignacio Zuloaga. The cross is of juniper wood, from trees chosen and cut down by General Franco himself. The image of Christ was placed on it and it was kept in the Chapel of the Pardo Royal Palace until the monument was completed.

◀ **68.** *The high altar.*

▼ **69.** *The main chapel: detail of the choir.*

70. *The main chapel: detail of one of the choir stalls.* ▶

71. *The chapel of the Most Holy: detail of the relief decorating the front of the altar.*

72. *The chapel of the Most Holy.*

73. The chapel of the Way of the Cross: detail of Christ recumbent.

74. The chapel of the Way of the Cross.

75. *The high altar: relief of the Burial of Christ on the front of the altar.*

76. *The high altar: detail of the Last Supper, on the back part of the altar.*

This image of Christ nailed to the cross is notable for the carefully-studied anatomic details as well as for the serenity emanating from the face, which shows no exaggerated signs of pain or suffering. Christ, His head high, raises His sight towards heaven, towards the cupola, seeking consolation and advice from God the Father. In this way both the cross and the cupola appear joined, united in a mutual, fluid communication. Once more serenity, sobriety and austerity issue from the monument, from its most magnificent works to its smallest details.

77. The high altar. Christ crucified, by Beovide.

THE CUPOLA

The culminating point of the transept comes, however, when the eyes are raised upwards to the extravagance of brightness and the colour of the mosaic in the cupola.

The art of mosaics, which has a long tradition in Spain, has often been used in works of a religious nature, since when thick pillars and walls are covered with countless mosaic tiles their materialness disappears, so that one seems to be within an ethereal, evanescent atmosphere with no visual limits, which is most appropriate for meditation and prayer. The believer becomes involved, becomes part of the whole and experiences an intense spiritual

78. *The cupola in the transept. Part of the mosaic — Christ the Lord.*

communication. In this way the immense cupola ceases to be a lineal representation and becomes a true vision of Glory.

The works of art within this magnificent mosaic should be carefully considered because of their great richness and variety. Outstanding is the figure of Christ the Lord, seated with a book in His left hand and His right hand offering a blessing; a circle of angels makes a halo around His head. Around Him are several groups of saints, heroes, martyrs, doctors of the Church and so on, on their way towards divine glory.

The Virgin, as opposed to Christ, is at prayer, the epitome of humility, as is customary in Christian painting, before the magnificence of her Son; she acts as inter-

79 and 80. *Details of the mosaic: groups of people at Christ's feet.*

81. *General view of the mosaic on the cupola.* ▶

82. *Detail of Christ.*

cessor of the people filing towards the Lord.

This work is by Santiago Padrós, who has positioned the groups extremely well, giving the impression of the cupola being higher and bigger than it really is. He used an obviously concentric design, the central part being occupied by Christ and the Virgin, His mother. In second place come the angels who together form the halos around Christ and the Virgin, and finally the various groups which in pyramidal design ascend from the outer section of the cupola towards the centre. The golden background gives greater scope to the subject and serves to separate the different hierarchies represented.

BUILDINGS AND WORKS ADJACENT TO THE MONUMENT

Behind the monument runs a portico measuring 300 metres long and 150 metres wide which leads to the buildings of the monastery, the choir school with a roll of 50 boys, and the hostel which can be used by the public (preferably for religious or cultural reasons), a centre for social studies and a library with thousands of copies.

Of particular interest in this area at the back of the monument is the magnificent, monumental bronze door, by Damián Villar González, from Salamanca. The door is made up of two main parts: a round arch and two jambs composed of twenty sections and a plinth. The relief work on the door shows the revelations made to St. John on the island of Patmos, where he wrote his *Apocalypse.* As has already been seen, this subject is also represented inside the basilica, that is on one of the tapestries in the Apocalypse series.

Benedictine monks are in charge of the religious services here and of the liturgal and cultural life in Los Caídos Valley. It

83. *The area behind the monument seen from the base of the cross.*

◀ **84.** *View of the rear entrance from the porticoes at the exit.*

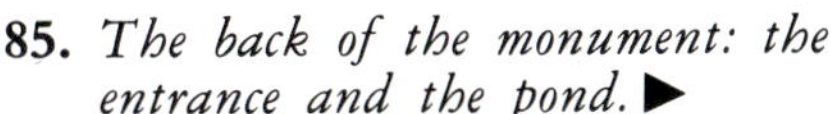

85. *The back of the monument: the entrance and the pond.* ▶

▼ **86.** *The back of the monument: side view of the porticoes.*

87. *The rear entrance: the bronze door.*

88. *The bronze door: detail of one of the sections.*

89. *The bronze door: detail of one of the sections.*

should be emphasized that they have always carried out their duties laudably.

A special reference should be made to the previously mentioned Centre for Social Studies. Set up in 1961, it organizes courses, conferences and seminaries of all types; its best known activity is the organization of round table discussions which are attended by national and foreign experts every year to debate many different subjects. These meetings give rise to editorial work with the publication of "The Social Studies Review".

Finally, laboratories for the control and study of electronic levels, the speed of the wind, gravimetry and the movement of the earth have been installed in the basements below the cross. The Department of Astronomy and Geodesy in Madrid University is in charge of them.

INDEX

INDEX

THE SPANISH NATIONAL HERITAGE COLLECTION

El Escorial
Los Caídos Valley
Las Huelgas Royal Monastery
The Royal Palace in Madrid
La Almudaina Palace